COSMIC WAVE THEORY

A NEW SCIENTIFIC TREATISE

BY

FRANCIS MLAGA

Contents

AUTHOR'S NOTE

Applications of the Cosmic Wave Theory in Astronomy.

Applications of the Cosmic Wave Theory in Fusion Energy Technology.

Philosophical Convictions on Nature, Science and Technology.

ABOUT THE AUTHOR

Only at eighteen, Francis Mlaga was awarded the 2006 National Academic Prize in Mathematics of Tanzania by the then president of the country Dr Jakaya Mrisho Kikwete. In 2007 Francis came to the United States and attended the State University of New York at Plattsburgh. Three years later he discovered a new equation for gravity in a wave perspective and founded the Cosmic Wave Theory. Apart from the aforementioned accomplishments this exceptional scientist is also renowned for his groundbreaking sci-fi/fantasy book series called Transdimensional Wars that he wrote and launched several years after writing and publishing his physics theory - Cosmic Wave Theory.

ACKNOWLEDGEMENTS

I would like to exponentially thank my late parents, Aloyce Mlaga
(a government mineral scientist) and Magreth Kyando (an entrepreneur) who passed on, in 2018 and 2019 respectively, for the fact that they saw, recognized and nurtured my academic talents ever since I was young to make who I am today and for bringing me to this dimension of existence so that I could manifest these phenomenal works and their wonders in science, mathematics, physics, technology and philosophy. Furthermore, I also present my heartfelt gratitude to my relatives, friends, colleagues, everybody who was involved in the process of publishing and marketing this book and readers for their support of such an intrinsic value. And last but not least, I would like to thank the almighty God for enabling me to write this monumental book in Physics and Astrophysics with practical applications in Astronomy and Fusion Energy Technology.

PRELUDE

Cosmic wave theory is a new physics theory based on the idea that the entire cosmos contains an ethereal wavy medium through which gravitational waves and subatomic particles such as photons of light, neutrinos and hydrogen molecules in fusion reaction travel. Furthermore cosmic wave theory contains a new equation for gravity in a wave perspective, progressing the works of Sir Isaac Newton and Albert Einstein.

This monumental work of science has been founded under the fundamental scientific methods of **Data Analysis, Observation, Hypothesis, Experimentation and Conclusion.**

INTRODUCTION

<u>Note</u>

On account of the mass of the earth, gravity is the most common force in daily life. It is one of the fundamental forces of nature; whereby the other three are strong nuclear force, electromagnetic force and weak nuclear force. Gravity is the weakest of these four forces of nature. For two protons in a nucleus the relative strength of gravity, weak nuclear force, electromagnetic force and strong nuclear force is 10^{-38}, 10^{-6}, 10^{-2} and 1 respectively.

<u>Pre-Newtonian Era</u>

About two millennia before Sir. Isaac Newton developed the law of universe l gravitation, Aristotle (384-322 B.C), an early Greek philosopher who considered causation as being totally inseparable to the substance of matter (matter as an entity with mass and volume) itself, in his theory of natural place and motion proposed that substances like a stone fell to the ground when thrown because it was seeking the center of the earth. Aristotle believed that matter in earth was made up of elements; earth, water, air and fire in solid , liquid, gaseous and plasma state, the concept known in an expanded form by all ancient Cosmogonies: Ancient Nile Valley Civilization, Ancient Chinese, and Ancient Hindus to describe the fundamental building blocks of all organic and psychic forms. He considered earth to be at the center encompassed by water, air and fire. And thus he was convinced that a stone fell when thrown because it was returning to its original position; the center of the earth. Aristotle's theory of natural place and motion lasted in the western world for almost two millennia.

During the 17th century a German Mathematician, Astrologer and Astronomer Johannes Kepler (1571-1630) who highly advanced Nicolas Copernicus' Heliocentric Theory emphasized that the sun emanates a motive force similar to magnetism and light to the plane of the planetary orbits. The sun's rotation on its axis accounts for the strengthening of the motive force. The strength of the motive force on a planetary body depends on its distance from the sun. He also held that this motive force governs the speed of planets around the sun. These concepts are embedded in his laws of planetary motion which are popularly known as Kepler's Laws of Planetary Motion. Kepler's first law of planetary motion states that, planets orbit the sun in elliptical paths with the sun at one focus. Kepler's second law of planetary motion states that, the closer a planet comes to the sun, the faster it moves. And Kepler's third law of planetary motion states that the ratio of the squares of the periods of any two planets revolving about the sun is equal to the ratio of the cubes of their mean distances from the sun.

Around the same period of time Galileo Galilee: (1564-1642) an Italian mathematician and Astronomer was unwittingly studying gravity. Contrary to Aristotle's conceptions of gravity, he showed that the acceleration of a falling object does not depend on its composition. That, in vacuum all objects fall at the same rate, regardless of their differences in mass. In air medium lighter objects fall slower than heavier objects due to air resistance. He established this constant rate on the surface of Earth to which falling objects accelerate as an average of 9.8 meters per second squared, which is now known as g (on the earth's surface.)

Newtonian Era

In 1666 at the age of 23, Isaac Newton (1642-1727) conceived of the most mysterious force - gravity unconventionally through observing the path of a falling. apple from one of the trees in his childhood home's garden in Woolsthorpe, England. In 1687 almost two decades after his serendipitous conception of gravity he published his worldly renowned scientific doctrine - "Philosophiae Naturalis Principia Mathematica."

In this masterpiece he set forth his law of universal gravitation, which states that every particle in the universe attracts every other particle with a force that is proportional to the product of their masses and inversely proportional to the square of the distance between them. Newton's "falling apple observation" accounts for the establishment of his Law, he is said to have been motivated and convinced: "If gravity acts at the tops of trees and even at the tops of mountains, then perhaps it acts all the way to the moon." He also established his three laws of motion which are popularly known as "Newton's Laws of Motion," with an aid of Kepler's Laws of planetary motion and Galileo Galilee's experimental conclusions concerning force and motion. Newton's first law of motion which is also famously known as the law of Inertia states that, everybody continues in its state of rest or of uniform velocity in a straight line unless exerted on by an external force. The second law states that the acceleration of an object is directly proportional to the force acting on it and is inversely proportional to its mass. And the third law states that, to every action there is an equal and opposite reaction. In spite of Newton's law of universal gravitation's controversy, in 1846 it led to the discovery of planet Neptune which is exerting a gravitational force on Uranus. Also Newton's law of universal gravitation made it possible to determine mass of the earth and the sun.

Post Newtonian Era

In 1855 Urbain Le Verrier (1811-1877) who largely contributed to the discovery of Neptune in 1846, studied the anomalies in mercury's orbit that are inconsistent with Newton's formula. By his detailed calculations he proposed that Mercury was being perturbed by the gravity of a smaller body orbiting even closer to the sun which leads to the slow constant drifting of the planet's orbit. This led to the popularity of the ideal that; an undiscovered planet called Vulcan existed in the intra-mercurial region. In 1915, Albert Einstein (1879-1955) published his general theory of relativity, a modification to Newtonian physics which is based on the idea that gravity is a consequence of space - time curvature (the curvature of a three dimensional space combined with a one dimensional time into a single manifold). And that all bodies with mass bend the space - time encompassing them. Einstein in his theory of general relativity further proposed that space – time around the sun was warped and thus this would lead Mercury to follow a closed ellipse when its orbit drew it proximity to the sun.

Post Einstein Era

In the 1930s scientists determined that the mass of the coma cluster was not enough to hold its galaxies in their orbits. Also they later observed the motions of galaxies in the clusters and inferred that the motions and rotations of galaxies in their clusters are not congenial with their masses. That, they perform their motions as if they had more significant mass than what can be observed. These phenomena led to the speculation of the existence of unusual, invisible or non-luminous matter called dark matter. In 1970 Vera Rubin discovered that the velocity of the Andromeda's outer regions is increasing and not decreasing, a property which supports the existence of dark matter. And further in 2006, after detecting the Bullet Cluster, scientists discovered that its distribution of visible matter, gas and invisible matter (or theoretical dark matter) is consistent with the dark matter hypothesis.

21st Century's Era

On this phenomenological journey of expanding our knowledge concerning the structure and behavior of the physical universe herein is a new developed theory of gravity - wave theory of gravity. It describes gravity and space in a transcendent model beyond what has already been presented by Isaac Newton and Albert Einstein.

The wave theory of gravity begins with redefining the concept of ether. The concept of ether can be considered as the most controversial scientific problem of the nineteenth century. It arose after the establishment of the theory of electromagnetism by James Clerk Maxwell (1831-1879).

Maxwell's equations predicted that the speed of light would be about 3.0×10^8 m/s within experimental error, which was later verified by Albert A. Michelson (1852-1931) using an eight sided rotating mirror a light source and a stationary mirror in a long evacuated tube. Maxwell's theory of electromagnetism does not specify a frame of reference from which the speed of light is. For instance in a set of phenomena whereby observers are traveling in a space craft at a speed of 1.8×10^8 m/s away from a source of light we might expect them to measure speed as the aforementioned speed of light in vacuum minus speed of the space craft but electromagnetism has no provision for relative velocity. It predicted the velocity of light to be the same as that in vacuum. This led the nineteenth century physicists to propose the existence of ether- a transparent medium that permeates all matter, which was considered by them to be at absolute rest. And thus they proposed that the velocity of light given by electromagnetism must be with respect to the ether. Nonetheless it seemed that electromagnetism was inconsistent with the relativity principle. Maxwell's equations were not the same in all initial frames of reference. They were the same only in a frame of reference where the speed of light was about 3.0×10^8 m/s, which appeared to the nineteenth century – physicists to be a frame of reference at absolute rest in the ether. In other frames of reference additional terms would have to be added to theoretically include the relative velocity.

In 1887 Albert Michelson and Edward Morley performed an experiment which is popularly known as the Michelson - Morley Experiment to measure the speed of light in different directions. They anticipated to find a variation depending on the orientation of their apparatus with respect to the ether, for they believed that just as a boat has different speeds with respect to the land when it moves downstream, upstream or across stream, so too light would have different speeds depending on the velocity of the ether past the earth. Nevertheless unwittingly they detected no variation at all. This led to the eruption of a variety of contradictory explanations to which none of them was accepted. Consequently the latter had led Albert Einstein in 1905 to introduce a special theory of relativity which contributed to a complete abandonment of the concept of ether in science. In 1915 he introduced a general theory of relativity. Despite the success that the general theory of relativity has in science it is said to be having some uncertainties.

It has been claimed that observations of DI Hercules, two young stars in an eclipsing binary system, made by Edwin Guinan in 1977 were not consistent with that which the general theory of relativity predicts. As the stars orbit around each other the axis of their orbit precesses at a rate of 1.05 degrees per century which is contrary to 4.27 degrees per century predicted by the general theory of relativity. The same inconsistence of the general theory of relativity is claimed to occur in another binary star system called AC Camelopardalis. While special relativity is partially consistent with the Wave Theory of Gravity and the whole of General Theory of Relativity may not be completely wrong, it is more probable that Einstein failed to unify gravity and electromagnetism to create a unified field theory due to a lack of knowledge concerning the ether.

With all due respect, evidently Einstein did not solve the ether problem nor did he attempt to approach it. Note that in science we are focusing on a quest for truth never should one avoid a scientific problem out of tact. Abandoning the main scientific problem to merely attempt to tackle its minor scientific problem is not technically solving the main scientific problem. Never can one find absolute truth about minor scientific problems without having a subtle knowledge of the main scientific problem itself. Einstein seems to many physicists to have solved mercury's anomaly problem.

However, the recent claimed uncertainties of the general theory of relativity and Einstein's deficiency in the exact knowledge of the ether should be a necessary force towards the reexamination of the cause of mercury's anomaly.

Unlike Einstein, Isaac Newton was aware of the ether, though he was seeking to have a full understanding of it as we can see in his book "Hypothesis" of 1675 he stated:

"Nature is a perpetual circulatory worker generating fluids out of solids, fixed things out of volatile and volatile
out of fixed, subtle out of gross and gross out of subtle, thus perhaps may all things be originated from ether."

So too was Thomas Young nineteenth century physicist who greatly contributed to the establishment of the wave theory of light, though he did not have a full grasp of the structural nature and behavior of it and endowed it with some properties that do not belong to it, as we can see in his scientific paper- "On the Theory of Light and Colors." he stated four hypotheses concerning light and the ether:

1. A luminiferous ether pervades the universe, rare and elastic in a high degree.
2. Undulations are excited in this ether whenever a body becomes luminous.
3. The sensation of different colors depends on the different frequency of vibrations excited by light in the retina
4. All material bodies have an attraction for the ethereal medium, by means of which it is accumulated within their substance, and for a small distance around them, in a state of greater density, but not of greater elasticity.

By philosophically and mathematically rethinking the ether problem in the twenty first century the Cosmic Wave Theory defines ether as a universal transparent medium of exceedingly high tenuity interpenetrating all matter and behaves in a wave aspect as water medium and gravity of a body in the ether medium as a resultant force of ether medium displaced by the body, which tends to return to its original position by acting at the center of the body. Most importantly Cosmic Wave Theory provides a new equation of gravity that is consistent in all systems.

Gravitational Force between two bodies, $F = \dfrac{n(m1 + m2)A^2V^2}{r^3}$

Whereby n is a constant refractive index of ether medium in which they are in, m1 and m2 is the masses, A is the gravitational amplitude of their resultant wave, V is the consequential velocity of the gravitationally influenced body and r is the mean distance between them from their centers.

Is the empty space really empty? Are there any similarities between the water medium and the ether medium? Can the empty space (ether) be displaced? How does gravity act at a distance? What is it about gravity that which makes it the weakest of the four forces of nature?

What is it about the wave principle of gravity that which makes it consistent in all systems? In a wave perspective how does the sun influence its planets gravitationally?

These are some of the most exponentially important questions about space (ether) and gravity of which answers have been revealed in this doctrine in a wave perspective.

Moreover, we shall see the differences between pure and impure bicorporal celestial systems. This first edition of the cosmic wave theory is mainly built by the wave principle of gravity. This new physics principle has been derived under the matrix of the scientific methods: Data Analysis, Observation, Hypothesis, Experimentation and Conclusion. And last but not least, as any other physics theory the cosmic wave theory has empirical evidence buried herein that's based on the harmony between it and Newton's law of universal gravitation in pure bicorporal celestial systems and the differences between it and the said Newton's Law, in impure bicorporal celestial systems. Indeed, this empirical evidence verifies it precisely.

Data Analysis

Fundamental Constants Used Herein

Gravitational Constant, $G = 6.67 \times 10^{-11} \text{ N m}^2/\text{kg}^2$

Speed of Light in the Ether, $C = 2.99792458 \times 10^8$ m/s

Mercury

Mass = 3.3×10^{23} kg

Radius = 2439 km or 2.439×10^6 m

 Mean Orbital Velocity = 48 km/s

Opposition at Aphelion = 70×10^6 km or 70×10^9 m

Opposition of Perihelion = 46.5×10^6 km or 46.5×10^9 m

Aphelion – Perihelion difference = 23.5×10^6 km or 23.5×10^9 m

Mercury – Sun Mean Distance = 58×10^6 km

Venus

Mass = 4.869×10^{24} kg

Radius = 6052 km or 6.052×10^6 m

 Mean Orbital Velocity = 35.02 km/s or 35020 m/s

Opposition at Aphelion = 108.9×10^6 km or 108.9×10^9 m

Opposition at perihelion = 107.5×10^6 km or 107.5×10^9 m

Aphelion - Perihelion difference 1.4×10^6 km or 1.4×10^9 m

Venus - Sun mean distance = 108.2×10^6 km or 108.2×10^9 m

Earth

Mass $= 5.97 \times 10^{24}$ Kg

Radius $= 6.38 \times 10^3$ km or 6.38×10^6 m

Mean orbital Velocity $= 29.8$ km/s or 29800 m/s

Opposition at aphelion = 152×10^6 km or 152×10^9 m

Opposition at perihelion = 147×10^6 km or 147×10^9 m

Aphelion - Perihelion Difference = 5×10^6 km or 5×10^9 m

Earth - Sun Mean Distance = 149.6×10^6 km or 149.6×10^9 m

Moon

Mass = 7.35×10^{22} kg

Radius = 1.738×10^3 km or 1.738×10^6 m
Mean orbital velocity = 1.022 km s or 1022 m/s
Opposition at apogee = 405508 km
Opposition at perigee = 363300 km
Apogee - Perigee Difference = 42208 km
Earth - Moon Mean Distance = 384404 km

Mars

Mass = 6.4 $\times 10^{23}$ kg

Radius = 3376.98 km or 3376.98×10^3 m
Mean Orbital Velocity = 24.1 km/s or 24100 m/s

Opposition at aphelion = 249.1×10^6 km or 249.1×10^9 m

Opposition at perihelion = 206.6×10^6 km or 206.6×10^9 m

Aphelion - Perihelion Difference = 42.5×10^6 km or 42.5×10^9 m

Mars - Sun Mean Distance = 227.976×10^6 km or 227.976×10^9 m

Jupiter

Mass = 1898.6×10^{24} kg
Radius" = 69911 km or 69911×10^3m
Mean orbital velocity = 13.07 km/s or 13070 m/s
Opposition at aphelion = 816.0×10^9m

Opposition at perihelion $= 740.6 \times 10^6$ km or 740.6×10^9 m

Aphelion - Perihelion Difference $= 75.4 \times 10^6$ km or 75.4×10^9 m

Jupiter - Sun mean Distance $= 778.4 \times 10^6$ km or 778.4×10^9 m

Saturn

Mass $= 568.46 \times 10^{24}$ kg
Radius $= 58232$ km or 58232×10^3 m
Mean orbital velocity $= 9.66$ km/s or 9660 m/s
Opposition at aphelion $= 1.5064 \times 10^9$ km

Opposition at perihelion $= 1.3476 \times 10^9$ km or 1.3476×10^{12} m

Aphelion -Perihelion Difference $= 158.8 \times 10^6$ km or $1\,58.8 \times 10^9$ m

Saturn – Sun Mean $= 1.4268 \times 10^9$ km
Distance

Uranus

Mass $= 86.83 \times 10^{24}$ kg
Radius $= 25362$ km or 25362×10^3 m
Mean Orbital Velocity $= 6.82$ km/s or 6820 m/s

Opposition at aphelion $= 3.0052 \times 10^9$ km or 3.0052×10^{12} m

Opposition at Perihelion $= 2.7340 \times 10^9$ km or 2.7340×10^{12} m

Aphelion-Perihelion Difference $= 271.2 \times 10^6$ km or 271.2×10^9 m Uranus – Sun Mean Distance $= 2.8710 \times 10^9$ km or 2.8710×10^{12} m

Neptune

Mass = 102.43×10^{24}kg

Radius = 24624km or 24624×10^3m

Mean Orbital Velocity = 5.48 km/s or 5480 m/s

Opposition at Aphelion = 4.5352×10^9km or 4.5352×10^{12}m

Opposition at Perihelion = 4.4580×10^9km or 4.4580×10^{12}m

Aphelion-Perihelion Difference= 77.2×10^6km or 77.2×10^9m

Neptune- Sun Mean Distance = 4.4983×10^9km or 4.4983×10^{12}m

Observation

On the Nature of the Outer Space.

It is quite known that in the physical universe matter exists in four different states-solid, liquid, gaseous and plasma state. And that matter in any state becomes a medium when it acts as a means of transmission of any kind of wave, particle, force or effect. For example air, water and diamond are some of the media that transmit light.

By observing the properties of air, water and diamond we can clearly see that all media have common properties'· and distinctive properties. Having an ability to allow the propagation of waves, particles, forces or effects and a potential to be displaced are some of the common properties, whilst density and the index of refraction are some of the distinctive properties. Explicitly, the first above mentioned common property is possessed by all media because without it they would not be so, and the second one is empirical because all media are forms of matter, which means they have mass and occupy space. A state of having a certain amount of mass and occupying space makes them have a potential to be displaced by other forms of matter. Density and index of refraction can be considered as some of the distinctive properties of media on account of their variation in all kinds of media, for example the density and index of refraction of water and air is 1000 kg/m^3 and 1.33, and 1.29 kg/m^3 and 1.0003 respectively.

Accordingly, when we observe the properties of the empty space we can clearly see that it exhibits a common property of being a means of transmission of elementary particles such as photons and neutrinos. Photons of light from the sun and distant star pass through the empty space to reach the earth and solar neutrinos emitted from the thermonuclear reactions of the sun pass through the empty space to reach the earth as well. Therefore this factor plus the fact that neutrinos interpenetrate matter, for example solar neutrinos interpenetrate the earth, strongly and clearly implies that there must be a universal transparent medium of exceedingly high tenuity that interpenetrates the entire structure of the physical universe through which these particles propagate.

Hypothesis

On the Architectonic and Behaviors of the Ether

In the preceding section we have seen that universal transparent medium which interpenetrates the whole structure of the physical universe and comprises the empty space or vacuum should be having exceedingly high tenuity. We know that light travels in the empty space at a constant speed of 2.99792458×10^8 m/s, and in air and water medium at a steady speed of 2.997025472×10^8 m/s and 2.254078632×10^8 m/s respectively. That light travels faster in the empty space than it does in air and water medium interprets that the medium which comprises the empty space has density that is less than that of air and water medium. It is thus extremely low density should be another property of the universal transparent medium that interpenetrates the whole of the physical universe.

These two properties of the universal transparent medium that interpenetrates the entire physical universe are congenial with the medium that Sir Isaac Newton, Christian Huygens, Thomas Young, and Sir William Thomson (Lord Kelvin) attempted to create a model of the entire ether medium. Therefore I am going to maintain the same name, though great modification concerning the whole of its architectonic and behavior shall be employed.

As we have discussed in the previous section all media are forms of matter and one of their common properties is having a potential to be displaced by other forms of matter. Accordingly, despite the proposition that the ether medium should be interpenetrating the whole structure of the physical universe, transparent and having extreme tenuity and exceedingly low density. It is a form of matter; matter in its extremely high and pervasive state and should be having a potential to be displaced by other forms of matter.

I conceive that the ether medium is as abundant in the physical universe as water medium in the earth and that the two media should characteristically relate to each other to a certain extent. We can imagine ether as water at an extremely high degree of· refinement or tenuity and pervasiveness. Thus any vibrations in the ether should cause ether waves, just like how water waves are formed due to vibrations in stagnating water.

And just as how water waves propagate to great distances at a certain fixed speed after being excited by vibrations in stagnating water, ether waves after being excited by vibrations in the ether medium should propagate to great distances at certain constant speed. And further, since the wave velocity of any wave depends on the properties of the medium in which it propagates, then the extreme tenuity and exceedingly low density of the ether medium should render its waves tremendously high and constant speed that should be equal to the speed of light.

Hence, by summarizing I now maintain that following hypotheses, which are commensurable with each other, concerning the architectonic and behavior of the ether.

The ether medium is universal The
ether medium is transparent.

The ether medium interpenetrates the entire structure of the physical universe.

The ether medium has an extremely high degree of tenuity

The ether medium has an exceedingly low density.

The ether medium behaves in a wave aspect as water medium.

The ether medium, has potential to be displaced by other forms of matter.

The ether medium when subjected to vibrations, allows the formation of ether waves, to which vibrations is their source, that travel at a speed of light constantly to large distances.

On the Nature of Gravity in a Wave Aspect

In a microscopic perspective we have seen that ether permeates all matter in the physical universe and it is material though subtle in nature. And that it has potential to be displaced by other forms of matter. This means all celestial bodies in the ether medium (space) displace various volumes of the ether medium with respect to their sizes just like an object which is fully submerged in water, it displaces a certain volume of water which is equal to its.

Apart from what we normally examine on earth, where everything is affected by its gravity: Suppose we have a spacecraft A in the ether. It will spontaneously displace a certain volume of ether which will be equal to its, just like how an object that is fully submerged in a water body such as a pond displaces a certain volume of water which is equal to its. The displaced ether will possess a certain resultant force as a reaction due to the action of displacing ether. The displaced ether with its resultant force will consequently be tending to return to its original position by acting at the center of spacecraft A. Without a thorough mathematical and philosophical reasoning one might think this resultant force, F is buoyant force, FB found by FB = PFgV, whereby, PF is the density of the fluid, g is the acceleration due to the earth's gravity, and V is the volume of the displaced fluid, or in other words with respect to the Archimedes' principle FB is equal to the weight of the displaced fluid. On the other hand the resultant force F.

Spacecraft A

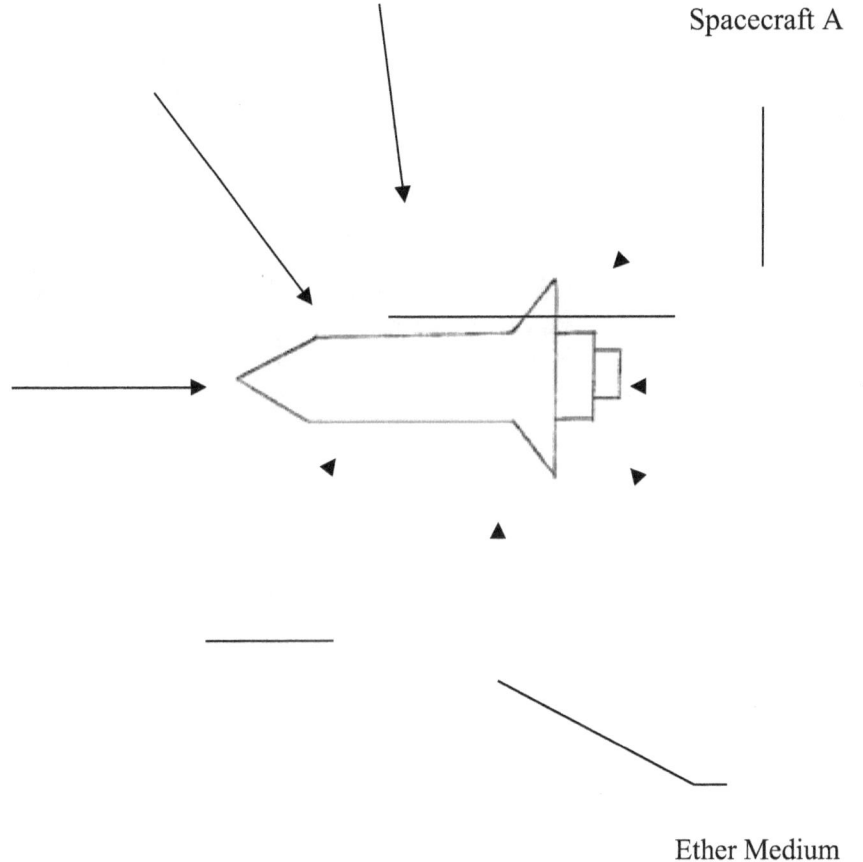

Ether Medium
displaced by Spacecraft A

Figure 3.1 A delineation to show Spacecraft A in the ether (space).

will not be equal to the buoyant force; FB since the displaced ether will be accelerating towards the center of spacecraft A, in an attempt to return to its original position. If there will beta celestial body close to it, it too will undergo the same process of displacing the ether, and. thus as a result the resultant forces of the two bodies will be pulling against each other.

Moreover, just like water medium, as a result of spacecraft A displacing the ether, ether waves will be formed, whereby their main source of vibrations is spacecraft A, of which mass is constant. These ether waves should also be traveling at or nearby the speed of light on account of the medium's exceedingly high tenuity and low density as we have discussed in the previous 0 section. Particles of the displaced ether will move by oscillating in simple harmonic motion, SHM as the ether waves, possessing the resultant force, F (force caused by spacecraft A), pass sinusoidally towards the center of spacecraft A. It follows that this resultant force should be gravity, since it is a force and consequence due to an object's presence in a medium and acts at the center of the object.

Gravity and Intensity Relation

We know that waves possess intensity. Accordingly, ether waves of the displaced ether should be possessing intensity as well. Intuitively just. like other forms of waves the intensity of the displaced ether should be directly proportional to: the mass of the source of vibrations in this case Spacecraft A.

From classical physics we know that: Intensity, I of a wave is the energy per unit time carried across unit area perpendicular to the direction of energy flow;

$$I = \frac{energy/\ time}{area}$$

Intensity, I of a sinusoidal wave is proportional to the square of the wave amplitude, A^2;

$$I \ \alpha \ A^2$$

And intensity, I of sinusoidal wave in a three dimensional medium of which source is constant decreases as the inverse square of the distance from the source, r^2

$$I \ \alpha \ \frac{1}{r^2}$$

It follows that, in our set of phenomena of spacecraft A, since energy is taken into consideration in a different manner of; the resultant force of the displaced ether medium acting at the center of spacecraft A, then the intensity of the waves of the ether medium displaced by spacecraft A will be defined as the force working at a distance per unit time carried across unit area perpendicular to the direction of the force:

$$I = \frac{\text{force} \cdot \text{distance/time}}{\text{area}}$$

Consequently as we can see, the intensity of waves of the ether medium displaced by spacecraft A will be proportional to the force of gravity of spacecraft A, and so too will this force of gravity be proportional to the energy of these waves, for sinusoidal wave intensity is proportional to sinusoidal wave energy.

Therefore on account -.of the relationship between intensity and gravity, it is thus force of gravity, F in sinusoidal ether waves (of which particles move in simple harmonic motion, SHM as the waves pass) is proportional to the square of the wave amplitudeA^2:

$$F \propto A^2$$

And further, this force of gravity, F in sinusoidal ether waves of displaced three dimensional ether medium of which source(spacecraft A) is constant in mass decreases as the inverse square of the distance from the source, r^2 due to its relationship with intensity:

$$F \propto \frac{1}{f^2}$$

Gravity and Mass Relation

We have seen that a force of gravity carried by a sinusoidal ether wave is directly proportional to the mass of the source of the wave (mass of the body in the ether). When two bodies with mass m1 and m2 respectively are in ether medium, the force of gravity, F between them is essentially a net vector sum of the force of gravity produced by the two bodies individually. It is thus the net gravitational force, F between the two bodies is accordingly directly proportional to the algebraic sum of

38

their masses, m 1 + m 2 (since mass is a scalar quantity):

$F \alpha \, m1 + m2$

Gravity and Acceleration Relation

It is merely common sense that the force of gravity between two bodies of which one is greater than the other in terms of mass is dominated by that due to the larger body. This means by determining the gravitational acceleration, a of the smaller body toward the larger body one is determining the effect of the net force of· gravity, F between them. It is therefore, the net force of gravity, F between the two bodies is directly proportional to the gravitational acceleration, a, of the smaller body:

$F \, \alpha \, a$

Experimentation

Wave Interference

Imagine that we have a new set of phenomena involving two objects P and Q, whereby P is larger than Q in terms of mass, dropped in a fixed water body such as a pond simultaneously. The two objects will form vibrations in the water body, which will lead to the formation of two sets of circular waves.

A constructive interference will be formed where crests of one wave recurring interfere crests of another wave and troughs interfere troughs in the same mode. And a destructive interference will occur where crests of one wave interfere troughs of another wave. At points between P and Q in a constructive interference, the resultant waves will consequentially oscillate up and down in a recurring manner with an amplitude larger than that of the waves formed by the vibrations of the two objects individually. Conversely, the points between the two objects here waves interfere with each other and become out of phase, will remain still.

Since object Q is smaller than P in terms of mass, the up and down oscillations of the resultant waves will cause it to have continuous horizontal displacements, whereby in the absence of experimental error the difference between the least and the greatest of these displacements will be equal to the amplitude of the resultant waves, provided no external force or object.

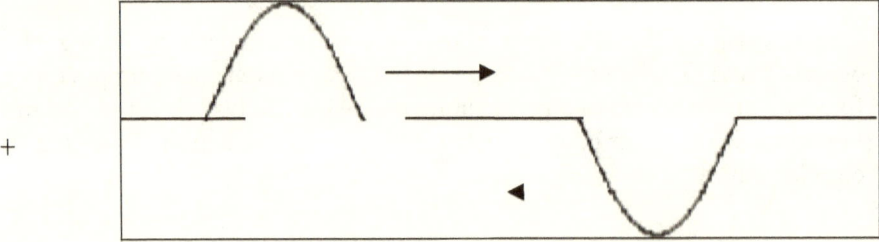

+

Two Waves approaching each other

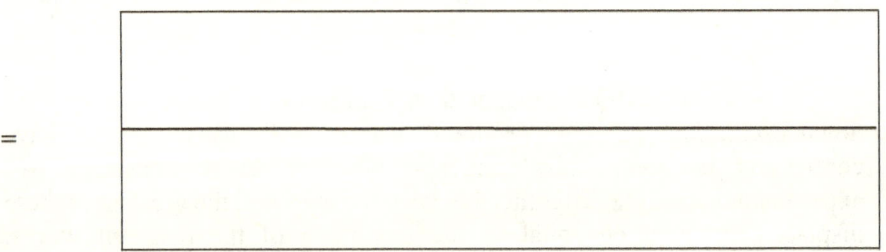

=

Their Resultant Wave

Figure 3.4

A delineation to show a destructive interference in which two waves
overlap precisely.

Sun   Earth   Moon

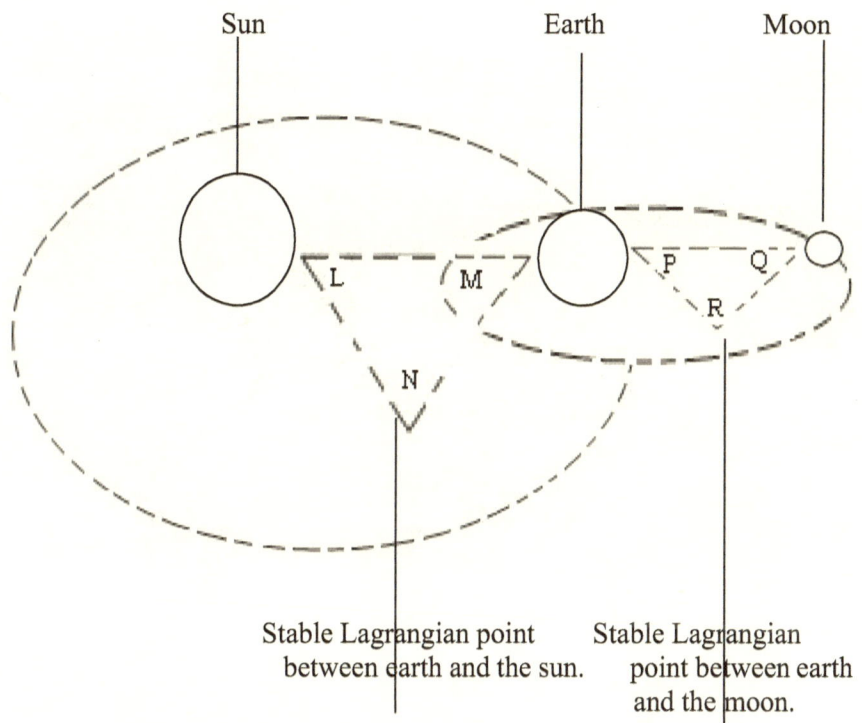

Stable Lagrangian point   Stable Lagrangian
between earth and the sun.   point between earth
                                  and the moon.

<u>Figure 3.5</u>

A delineation to show stable lagrangian points between Earth and the Moon and the Sun where smaller bodies tend to remain at rest. $< L = < M = < N$ and $< P = < Q = < R$

Figure 3.2

A delineation to show two crests of waves due to object P and Q with amplitude Ap and Aq respectively, approaching each other.

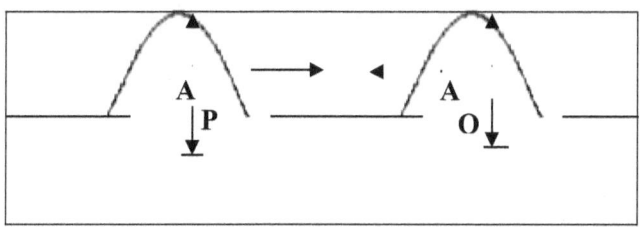

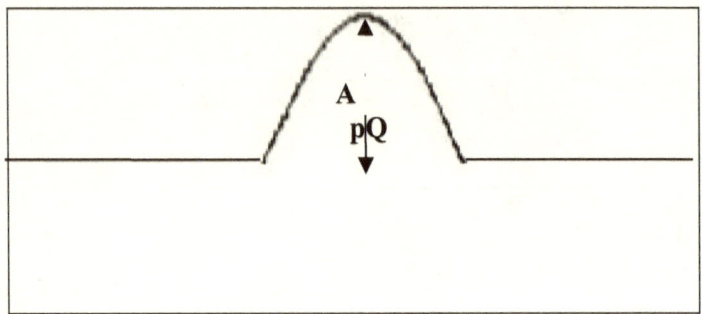

Figure 3.3

A delineation to show the amplitude of the resultant wave, ApQ formed after a constructive interference of wave crest due to object P and Q.

will be applied or added to the system in the region between P and Q. And according to the principle of superposition, in the region where two waves interfere with each other the resultant displacement is the algebraic sum of their separate displacements whereby an amplitude or displacement of a crest is considered positive and that of a trough negative.

The same set of phenomena will occur when two spacecrafts L and M of significantly large mass such that L is greater than M in mass, placed in space(ether: whereby the region between them is unoccupied. Both objects will separately displace ether and thus consequently the displaced ether will tend to return to its original position with a force of gravity carried by its waves exerting at the centers of the objects. The ether displaced by the two objects will make two set: of ether waves. A constructive interference will be formed where crests of one ether wave continuously interfere crests of another ether wave and troughs interfere troughs in the same manner. A destructive interference will be formed where crests of one ether wave interfere troughs of another ether wave. At points between L and M (in a constructive interference), where two waves produced by the two bodies separately interfere with each other, the resultant wave will form oscillations up and down in a continuation with an amplitude larger than that of the wave produced by the two bodies individually. Conversely, the points between the two objects where waves interfere with each other and become out of phase, will remain still. Since spacecraft M is smaller than L in terms of mass, the up and down oscillations of the resultant ether waves in constructive interferences will cause it to have continuous horizontal displacements, whereby in the absence of experimental error the difference between the least and greatest displacement will be equal to the amplitude of the greatest resultant wave provided no external force or object will be added to the system in the region between L and M. The points between L and M where ether waves due to L with those due to M interfere with each other destructively will form Lagrangian points whereby stable Lagrangian points between L and M will be formed where crests and troughs of ether waves produced by the two objects overlap precisely and remain still, as illustrated in figure 3.4. Lagrangian point is a point in space at which a smaller body under the gravitational influence of two larger ones will remain approximately at rest relative to them. These particular points in the ether exist in bicorporal celestial systems of massive bodies such as the\ earth-moon system. In these systems five theoretical Lagrangian points tend to exist of which only two are stable. The stable Lagrangian points retain smaller bodies regardless of the slight perturbations caused by external gravitational influences produced by the larger ones. In other words, at stable Lagrangian points gravitational forces are precisely balanced and hence smaller bodies at the points remain approximately at rest. Each stable Lagrangian point forms one vertex of an equilateral triangle and other two vertices are formed by the two larger bodies as shown in figure 3.5.

Conclusion

Wave Principle of Gravity

From the preceding section we can see that there are four properties of the force of gravity between two bodies with mass. These properties are:

1. Force of gravity between two bodies, F is proportional to the square of the amplitude, A^2 of their resultant ether wave:
   $$F \alpha A^2$$

2. Force of gravity between two bodies, F is inversely proportional to the square of the distance between them, r^2 :
   $$F \alpha \frac{1}{r^2}$$

3. Force of gravity between two bodies, F is directly proportional to the gravitational acceleration, a of the influenced body:
   $$F \alpha a$$

4. Force of gravity between two bodies, F is proportional to the algebraic sum of their masses, m1 + m2:
   $$F \alpha m1+m2$$

From above the compound proportionality of F, A, m1, m2, a and r is:

$$F \alpha \frac{(m1 + m2)A^2 a}{r^2}$$

$$F = \frac{K(m1 + m2)A^2 a}{r^2}$$

Since the refractive index, n of each medium is proportional to its density and is constant, then it becomes the constant, K of the equation above thus:

$$F = n\,(m1 + m2)A^2 a/r^2$$

But n = 1.0000 in the case of the ether medium, therefore:

$$F = \frac{(m1 + m2)A^2 a}{r^2}$$

Hence we can now state a principle of gravity which is based on the ideal form of wave in accordance with the properties of gravity hitherto inferred, that: The force of gravity, F between two bodies in a medium is directly proportional to the sum of their masses, $m_1 + m_2$, the square of the gravitational amplitude of their resultant wave, A^2 and the gravitational acceleration, a of the influenced body but inversely proportional to the square of the mean distance between them from their centers, r^2 for example the force of gravity between two bodies in the ether medium(space) is found by:

<u>A New Formula;</u>

Force of Gravity, F
$$=(m1 + m2)\,A^2\,a/r^2$$

In all celestial systems composed by two celestial bodies whereby one of which is orbiting the other with a uniform velocity, it is known that the acceleration of the revolving body(influenced body) is considered a centripetal acceleration, aR which acts toward the influencing body at the center of the influencing body's orbit, as shown in figure 3.6.

The magnitude of the centripetal acceleration of the influenced body with the respect to the influencing body is

$$aR = \frac{v^2}{r}$$ whereby v is the velocity of the

Influenced body of r is the mean distance between them from their centers r.

From the wave principle of gravity the gravitational force between two bodies F in the ether (space) is determined by:

$$F = \frac{(m1+m2)A^2a}{r^2}$$ whereby m1 and m2 are their masses, A is

the gravitational amplitude of their resultant wave, a is the gravitational acceleration of the influenced body and r is the mean distance between them from their centers. This means in a set of phenomena in which the influenced body is revolving around the influencing body, its centripetal acceleration, aR is the gravitational acceleration a by substituting aR in the wave principle of gravity it

$$F = \frac{n(m1 + m2)A^2 aR}{r^2}$$

follows that gravitational force,

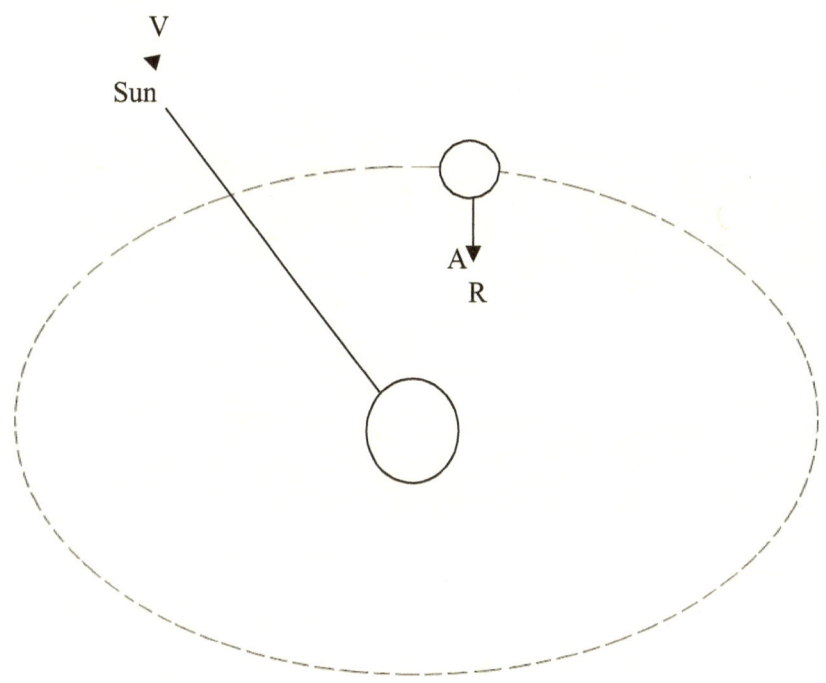

Figure 3.6

A delineation to show the centripetal acceleration of the earth towards the sun, aR.

A Further Analysis of Gravity and the Ether

In its transcendental meaning, we have seen that gravity of a body in any medium is the resultant force of the medium displaced by it, which tends to return to its original position by acting at the very center of the body. For celestial bodies in the ether medium (free space), for instance the earth, the same behavior of nature holds.

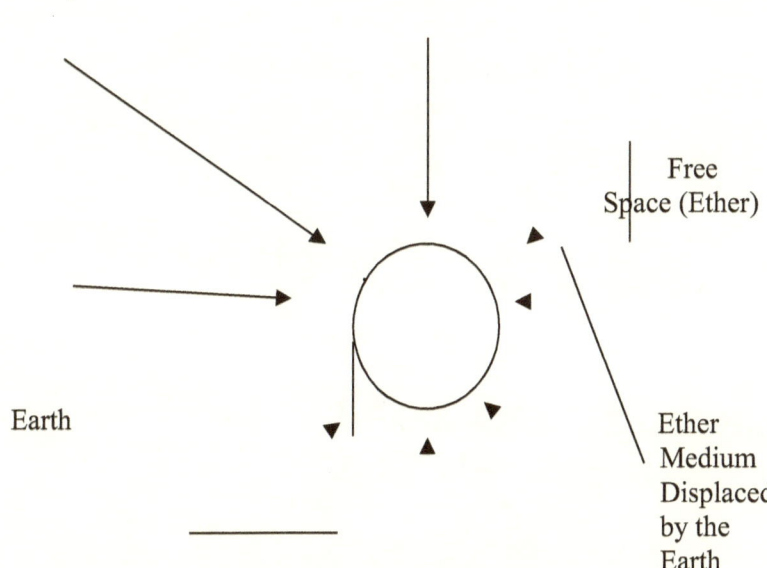

Earth

Free
Space (Ether)

Ether
Medium
Displaced
by the
Earth

<u>Figure 3.7</u>

A delineation to show Earth in the ether medium and the ether medium displaced by it tending to return to its original position by acting at the center of the earth.

$$aR = V^2/r \quad \text{then}$$

Gravitational Force, $\quad F = (m1 + m2)A^2 \; \dfrac{\frac{V^2}{r}}{r^2}$

Gravitational Force, $F = \dfrac{(m1 + m2)A^2 V^2}{r^3}$

Whereby m1 and m2 are their masses, A is the gravitational: amplitude of their resultant ether wave, V is the consequential velocity of the influenced body and r is mean distance between the two bodies from their centers.

Accordingly, in general the force of gravity, F between two bodies in a medium whereby one is orbiting the other is determined by:

$$\text{Gravitational force, } F = \underline{n}\,\dfrac{(m1 + m2)A^2 V^2}{r^3}$$

Whereby n is the refractive index of the medium in which they are in, m1 and m2 are their masses, A is the gravitational amplitude of their resultant wave, V is the consequential velocity of the influenced body and r is the mean distance between them from their centers.

<u>Figure 3.8</u>

A delineation to show the gravitational influence between earth and the
moon due to the ether medium displaced by them.

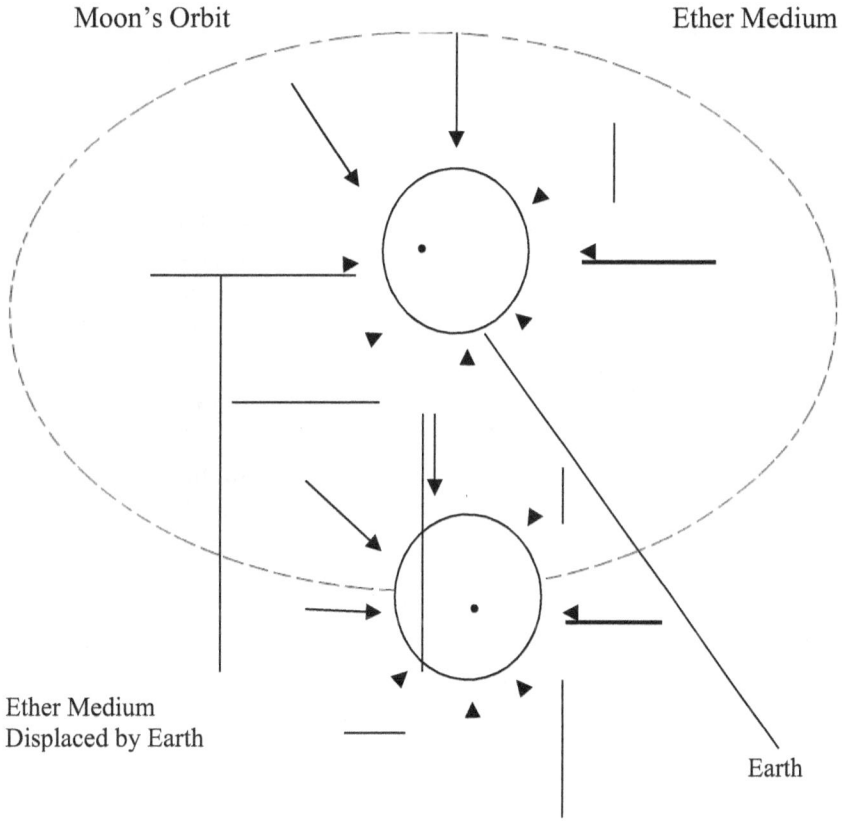

Moon's Orbit   Ether Medium

Ether Medium
Displaced by Earth

Earth

Ether Medium Displaced by
the Moon

The earth constantly displaces a volume of the ether medium of 1.08×10^{21} m^3 that equals its volume. As a reaction the action of displacing the ether, the displaced ether attains a resultant force that is possessed by waves of the displaced ether and acts toward the center of the earth, as the displaced ether attempts to return to its original position by acting at the center of the earth. This resultant force is the force of gravity of the earth since it is a force and consequence due to the presence of the earth in space (ether medium), and its constant action of displacing the ether. Consequently, each object on the earth's surface is influenced by the resultant force(force of gravity of the earth)carried by waves of the ether medium displaced by the earth that attempts to return to its original position by acting at the earth's center.. This is why everything on the earth's surface accelerates toward the earth's center.

For bodies in space that are within: or nearby the region covered by the ether medium displaced by the earth will also be influenced by the resultant force of the medium displaced the earth, and thus accelerate toward its center. A perfect natural example in this case is the earth's natural satellite the moon. However by using the moon as an example, due to its possessing mass it also displaces the ether medium in space. The moon constantly displaces a volume of the
either of 21.99×10^{18}m^3 that equals its volume. In the same manner as the earth its displaced ether attains a resultant force carried by the waves of the displaced ether toward the center of the moon as the displaced ether itself attempts to return to its original position by acting at the center of the moon. This resultant force is the moon's force of gravity. It follows that as a result of the pulling of the force of gravity of the earth against that of the moon, the moon orbits about the earth, since its gravity is smaller than that of planet earth. The same phenomenon occurs between stars and their respective planets.

The sun constantly displaces a volume of the ether of approximately 1.4×10^{2}m^3 equal to its volume. As a result of the action of displacing the ether medium, in reaction the displaced ether attempts to return to its original position by acting at the center of the sun with a resultant force possessed by the waves of the displaced ether. This resultant force is the force of gravity of the sun due to the sun's presence in the ether medium(space) and its constant action of displacing the ether. Just like the earth and the moon, since the sun contains ether, then the displaced ether with its waves carrying the force of gravity of the sun, tends to compress the ether medium that is already a constituent of the sun, at a distance toward the center of the sun. It follows that, the solar planets which are within the region covered by the ether medium displaced the sun, or nearby it, tend to be influenced by the force of gravity carried by the waves of the ether medium displaced by the sun which consequentially acts toward the center of: the sun as an attempt to return to its original position. However, on account of the planets masses they independently also have forces of gravity carried by waves of the ether medium displaced by them, which similarly act at their centers. Hence as a result of the pulling of their gravitational forces against that of the sun separately, they orbit about the sun, since their individual gravitational forces are less than the sun's.

The same behavior of nature holds to all celestial bodies in space in its entirety, since the ether medium pervades the whole of the physical universe (and is a mere constituent of all celestial bodies). Moreover, we can see that due to gravity being a resultant force of a medium displaced by a body carried by waves of the medium displaced by the body toward the center of the body, as the displaced medium attempts to return to its original position by acting at the center of the body, and that it depends on the body's mass, then it is a weak force for bodies of small masses and thus the weakest of all four forces of nature. It is effective only for bodies of great masses such as celestial bodies.

3.3.3 On the Pure Bicorporal Celestial Systems

For a system of two bodies in which the region between them is free of other bodies the gravitational amplitude of their resultant wave must be equal to the change in displacement of the influenced body. Accordingly, in systems of stars and their planets in which the region between them is free of other celestial bodies, the gravitational amplitude of the resultant wave due to them must be equal to the aphelion and perihelion- difference of their respective planets. And similarly, for celestial systems of planets and their natural satellites in which the region between them is not occupied by other celestial bodies, like the earth-moon system the gravitational amplitude of their resultant wave must be equal to the change in the displacements of their respective natural satellites.

In figure 3.9, 3.10, 3.11, ether waves produced by the moon and earth interfere mostly constructively at each position creating a resultant wave which varies with respect to the position of the moon from earth. The closer the moon is to earth the greater their resultant wave and the farther the moon is from earth the less greater their resultant wave. By supposing that the moon is at its closest position to earth at point D and furthest position from earth at point K, it follows that the difference between the greatest displacement of the moon from earth (apogee) at point K and the least displacement of the moon from earth (perigee) at point D, must be equal to the gravitational amplitude of the resultant ether wave due to earth and moon.

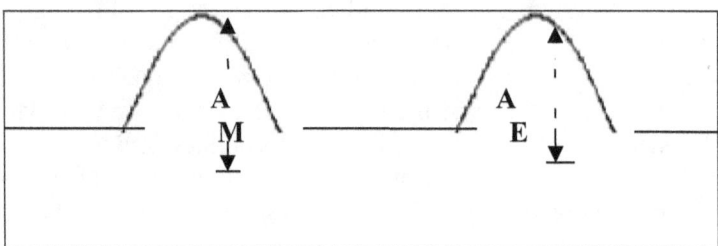

Figure 3.9

A delineation to show the amplitudes of ether waves formed by the moon and earth AM and AE respectively.

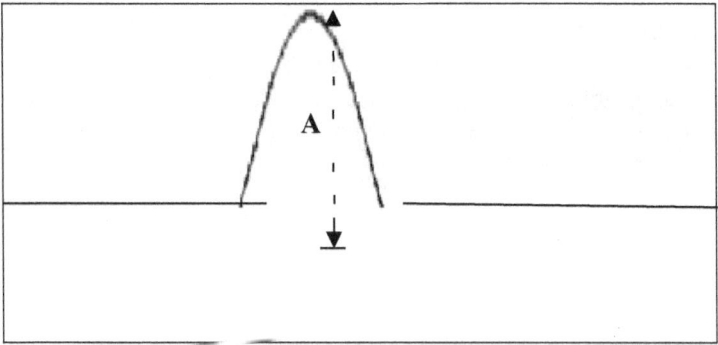

Figure 3.10

A delineation to show the gravitational amplitude of the resultant ether wave formed by the constructive interference of ether waves formed by the moon and earth.

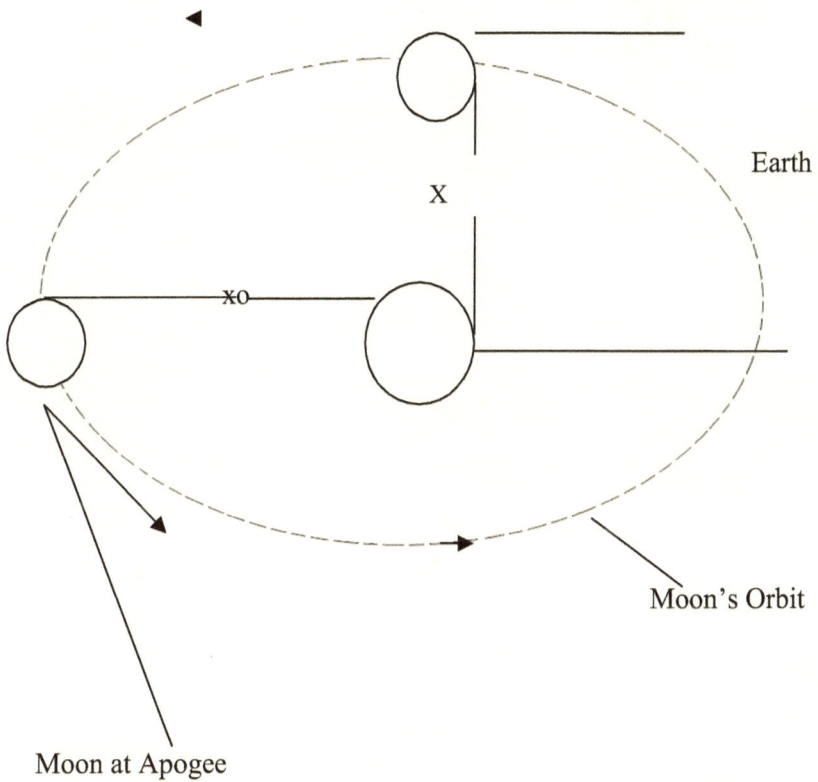

Moon at perigee

Earth

X

xo

Moon's Orbit

Moon at Apogee

Figure 3.11

A delineation to show the moon at the perigee and apogee whereby the apogee-perigee difference is equal to the gravitational amplitude, A of the resultant ether wave produced due to the gravitational interaction of the earth and moon, $A = x_o - x$.

At this point since we know the equation to determine the net gravitational force between two bodies then let us observe the phenomenal relationship between the wave principle of gravity and Newton's law of universal gravitation in one of pure solar bicorporal celestial systems, the earth-moon system. Pure bicorporal celestial system is a system of two celestial bodies in which the region between them is totally free of other celestial bodies. Relatively pure solar bicorporal celestial system is a system of two celestial bodies whereby one of which is the sun and the region between them is free of other celestial bodies.

Planetary, one of bicorporal systems that make pure bicorporal celestial systems are systems of planets and their closest natural satellites for example the earth-moon and so forth.

Gravitational Force between Earth and the Moon with respect to Newton's Law of Universal Gravitation.

According to our data analysis mass of earth, m1 is 5.97×10^{24} kg, mass of the moon, m2 is 7.35×10^{22} kg, mean distance between them from their centers, r is 384404000 m and the gravitational constant, G is 6.67×10^{-11} N.m²/kg², therefore gravitational force between earth and moon, F is:

$$\text{Gravitational Force, } F = G\frac{m1\,m2}{r^2}$$

$$= \frac{6.67 \times 10^{-11} \times 7.35 \times 10^{22} \times 5.97 \times 10^{24}}{(384404000)^2}$$

$$= \frac{292.676265 \times 10^{35}}{1.477664352 \times 10^{17}}$$

$$F = 1.980668104 \times 10^{20} \text{ N}$$

$$F = 1.98 \times 10^{20} \text{ N}$$

Thus according to the law of universal gravitation the gravitational force between earth and the moon is about 1.98×10^{20} N.

Gravitational Force between Earth and the Moon with respect to the Wave Principle of Gravity.

From our data analysis we know that mass of the earth is 5.97×10^{24} kg, mass of moon m2 is 7.35×10^{22} kg, mean distance between them from their centers, r is 38440400 m, apogee-perigee difference, A is 4220800 m and moon's orbital velocity, V is 1022m/s, therefore gravitational force between earth and the moon, F is:

$$\text{Gravitational Force, } F = n \frac{(m1 + m2)A^2V^2}{r^3}$$

$$= \frac{1(5.97 \times 10^{24} + 7.35 \times 10^{22})(42208000)^2(1022)^2}{(384404000)^3}$$

$$= \frac{10^{22}(597 + 7.35) \times 1.781515264 \times 10^{15} \times 1044484}{5.680200876 \times 10^{25}}$$

$$F = 1.9797765290 \times 10^{20} \text{ N}$$

$$F = 1.98 \times 10^{20} \text{ N}$$

According to the wave principle of gravity the force of gravity between earth and the moon is 1.98×10^{20}N. From our calculation we can see that the wave principle of gravity and the law of universal gravitation, when applied in the earth-moon system, they give results which are equal. This means that wave principle of gravity and the law of universal gravitation are absolutely consistent with each other in this bicorporal celestial system.

The consistency of wave principle of gravity and Newton's law of universal gravitation in the earth-moon system is a monumental phenomenon which implies that the wave principle of gravity and the law of universal gravitation are consistent with each other in all pure bicorporal celestial systems due to the absence of external gravitational forces from the region between the two celestial bodies.

3.6 On the Impure Bicorporal Celestial Systems

Impure bicorporal celestial systems are systems composed by two celestial bodies in which the region between them is consisting of another celestial body or bodies. Impure celestial systems exist between planets and their respective natural satellites (except the planets' closest satellites), stars and their respective planets, galactic centers of galaxies with black holes and their respective stars, and so forth.

By using the wave principle of gravity and the law of universal gravitation we are going to verify the existence of impure solar bicorporal celestial systems. Impure solar bicorporal celestial systems are impure bicorporal celestial systems that exist in the solar system in which the sun is the influencing body. The following calculations are for determining the state of the bicorporal celestial systems of the sun and its respective planets.

On the Venus- Sun System

From astronomy, Venus is a second planet from the sun and Mercury lies on the region between Venus and the sun. By taking the ether vortex that theoretically exist in the intra-mercurial region into consideration due to massive variation between the universal law of gravitation and the wave principle of gravity, the Venus-sun system is affected by the gravitational force produced by it. This means in the Venus-sun system Venus is exerted on by the gravitational force produced by the sun, the ether vortex in the intra-mercurial region and Mercury. The wave principle of gravity will determine the net gravitational force acting on Venus in the Venus-sun system whereas the law of universal gravitation will give us the force of gravity between Venus and the sun as if there are no celestial bodies between them. Therefore the variation between wave principle of gravity and the law of universal gravitation will verify that the Venus-sun system is an impure solar bicorporal celestial system.

Force of Gravity between Venus and the Sun with respect to the Wave Principle of Gravity.

From our data analysis mass of the sun, $m\,1$ is 1.99 x 1030 kg, mass of Venus, $m\,2$ is 4.869 x 1024 kg mean distance between them from their centers, r is 108.2 x 10^9 m, aphelion - perihelion difference of Venus, A is 1.4 x 10^9 m and mean orbital velocity of Venus, v is 35020 m/s, therefore the gravitational force between Venus and the sun, F is: By the Wave Principle of Gravity:

$$\text{Gravitational force, } F = \frac{n(m\,1 + m\,2)A^2v^2}{r^3}$$

$$F = \frac{1(1.99 \times 10^{30} + 4.869 \times 10^{24})(1.4 \times 10^9)^2(35020)^2}{(108.2 \times 10^9)^3}$$

$$F = \frac{10^{24}(1990000 + 4.869) \times 1.96 \times 10^{18} \times 1226400400}{1266723.368 \times 10^{27}}$$

$$F = \frac{4.783463824 \times 10^{57}}{1.266723368 \times 10^{33}} \, F$$

$$= 3.775249768 \times 10^{24} N \, F =$$

$$3.78 \times 10^{24} N$$

Thus by the wave principle of gravity the force of gravity between Venus and Sun is 3.78×10^{24} N

Gravitational Force between Venus and the Sun with respect to the Law of Universal Gravitation

From our data analysis we know that mass of the sun, m1 is 1.99×10^{30} kg, mass of Venus, m2 is 4.869×10^{24} kg, mean distance between Venus and the Sun from their centers, r is 108.2×10^{9} m and the gravitational constant G is 6.67×10^{-11} $N.M^2/kg^2$. It is thus the gravitational force between Venus and the sun, F is by the Law of universal gravitation,

$$\text{Gravitational force, } F = G \frac{m1m2}{R^2}$$

$$F = \frac{6.67 \times 10^{-11} \times 1.99 \times 10^{30} \times 4.869 \times 10^{24}}{(108.2 \times 10^{9})^2}$$

$$F = \frac{6.46276977 \times 10^{44}}{1.170724 \times 10^{22}}$$

$$F = 5.520318854 \times 10^{22} N$$

$$F = 5.52 \times 10^{22} N$$

Hence by the law of universal gravitation the gravitational force between Venus and the sun is about $5.52 \times 10^{22} N$.

On the Earth – Sun System

In the Earth Sun System, earth as a third planet from the sun;
Mercury the ether vortex in the intra-mercurial region and

Venus exist between them. The wave principle of gravity on the Earth - Sun system will give us the net gravitational force acting on earth (in its system with the sun), whereas the law of universal gravitation will give us the force of gravity between earth and the sun as if the region between them is not occupied by other celestial bodies. Therefore variation between the wave principle of gravity and the law of universal gravitation in the earth - sun system will verify that it is an impure solar bicorporal celestial system.

Gravitational Force between earth and the Sun with respect to the Wave Principle of Gravity.

According to our data analysis we know that mass of the sun, m1 is 1.99×10^{30} kg, mass of earth, 5.97×10^{24}kg, mean distance between them from their centers, r is 149.6×10^9 m, aphelion and perihelion difference of the earth, A is 5.0×10^9 m and the mean orbital velocity of earth , v is 29800 m/s. Therefore the gravitational force between earth and the sun, F is:

Under the wave principle of gravity;

$$\text{Force of Gravity, } F = n \, (m1 + m2) \frac{A^2 v^2}{r^3}$$

$$F = \frac{1(1.99 \times 10^{30} + 5.97 \times 10^{24})(5 \times 10^9)^2 (29800)^2}{(149.6 \times 10^9)^3}$$

$$F = \frac{(1990000 + 5.97)10^{24} \times (5 \times 10^9)^2 (29800)^2}{3348071.936 \times 10^{27}}$$

$$F = 1.319569095 \times 10^{25}N$$

$$F = 1.32 \times 10^{25}N$$

Therefore by the wave principle of gravity the gravitational force between earth and the sun is about $1.32 \times 10^{25}N$.

Gravitational Force between Earth and the Sun with respect to the Law of Universal Gravitation.

According to our date analysis we know that mass of the sun, m1 is 1.99×10^{30}kg, mass of earth, m2 5.97×10^{24}kg, mean distance between earth and the sun from their centers, r is meters 149.6×10^{9} and the gravitational constant; G is 6.67×10^{-11}N. m^2/kg^2, thus the gravitational force between earth and the sun, F is

Under the Law of universal gravitation;

Gravitational force, $F = G \dfrac{m1 m2}{r^2}$

$= 6.67 \times 10^{-11} \times 1.99 \times 10^{30} \times 5.97 \times 10^{24}/$
$(149.6 \times 10^{9})^2$

$= 79.241601 \times 10^{43}/$
22.38016×10^{21}

$= 3.540707528 \times 10^{22} N$

$F = 3.54 \times 10^{22} N$

Thus by the Law of universal gravitation the gravitational force between earth and the sun is about $3.54 \times 10^{22} N$.

On the Mars-Sun System

As a fourth planet from the sun, mars in its bicorporal system with the sun is acted on by external gravitational forces produced by Mercury, the ether vortex in the intra-mercurial region, Venus and the Earth, (other than that generated by the sun). The wave principle of gravity will determine the net gravitational force acting on Mars in its system with the sun. The law of universal gravitation will give us the gravitational force between mars and the sun as if the mars – sun system is a pure bicorporal celestial system. The variation between the wave principle of gravity and the law of universal gravitation on the mars-sun system will verify that it is an   impure   solar   bicorporal   celestial   system

Gravitational Force Between Mars and the Sun with Respect to the Wave Principle of Gravity

According to our data analysis the mass of the sun, m1 is 1.99×10^{30} kg, mass of mars, m2 is 6.42×10^{23} kg mean distance between them from their centers, r is 227.976×10^{9} m, change in displacement of mars (aphelion-perihelion difference), A is 42.5×10^{9} m and the mean orbital velocity, V of mars about the sun is 24100 m/s. thus the gravitational force between mars and the sun, F is:

$$\text{Gravitational force,} \quad F = n\,(m1 + m2)\frac{A^{2}V^{2}}{r^{3}}$$

$$= \frac{(1.99 \times 10^{30} + 6.42 \times 10^{23})(42.5 \times 10^{9})^{2}(24100)^{2}}{(227.976 \times 10^{9})^{3}}$$

$$= \frac{10^{24}(1990000 + 0.642) \times 1806.25 \times 10^{18} \times 580810000}{11848609.55 \times 10^{27}}$$

$$F = 1.761967013 \times 10^{26} N$$

$$F = 1.76 \times 10^{26} N$$

Therefore by the wave principle of gravity the gravitational force between mars and the sun is about 1.76×10^{26} N.

Gravitational Force between Mars and the Sun with respect to the Law of Universal Gravitation.

From our data analysis we know that mass of the sun m1 is 1.99×10^{30} kg, mass of mars, m2 is 6.42×10^{23} kg mean distance between mars and the sun from their centers, r is 227.976×10^{9} m and the gravitational constant G is 6.67×10^{-11} N m^2/kg^2 therefore the gravitational force between mars and the sun, F is: by applying the Law of universal gravitation;

$$\text{Gravitational force, } F = \frac{G \, m1 \, m2}{r^2}$$

$$= \frac{6.67 \times 10^{-11} \times 1.99 \times 10^{30} \times 6.42 \times 10^{23}}{(227.976 \times 10^{9})^2}$$

$$= \frac{85.214586 \times 10^{42}}{51.97305658 \times 10^{21}}$$

$$F = 1.639591581 \times 10^{21} \text{N}$$

$$F = 1.64 \times 10^{21} \text{N}$$

Therefore by the law of universal gravitation the gravitational force between mars and the sun is about 1.64×10^{21} N

On the Jupiter-Sun System

As a fifth planet from the sun, Jupiter in its bicorporal system with the sun is acted on by external gravitational forces produced by mercury, the ether vortex in the intra-mercurial region, Venus, Earth and Mars. The wave principle of gravity will determine the next gravitational force acting on Jupiter in its system with the sun. The law of universal gravitation will give us the gravitational force between Jupiter and the sun as if the jupiter – sun system is a pure bicorporal celestial system. The variation between the wave principle of gravity and the law of the universal gravitation on the Jupiter-sun system will verify that it is an impure solar bicorporal celestial system.

Force of Gravity between Jupiter and the Sun with respect to the Wave Principle of Gravity

From our data analysis we know that mass of the sun m1 is 1.99×10^{30} kg, mass of Jupiter, m2 is 1898.6×10^{24} kg mean distance between Jupiter and the sun from their centers, r is 778.4×10^{9} m, Jupiter's aphelion-perihelion difference, A is 75.4×10^{9} m and the mean orbital velocity of Jupiter V is 13070m/s. Therefore the mean the force of gravity between Jupiter and the sun, F is: Under the wave principle of gravity;

Force Gravity, $F = n \dfrac{(m1 + m2)A^2V^2}{r^3}$

$$= \frac{1(1.99 \times 10^{30} + 1898.6 \times 10^{24})(75.4 \times 10^{9})^2(13070)^2}{(778.4 \times 10^{9})^3}$$

$$= \frac{10^{24}(1990000 + 1898.6) \times 5685.16 \times 10^{18} \times 170824900}{47163766.3 \times 10^{27}}$$

$$= \frac{19.34465966.58 \times 10^{61}}{4.71637663 \times 10^{35}}$$

$F = 4.101593457 \times 10^{26}N$

$F = 4.10 \times 10^{26}N$

By the wave principle of gravity the force of gravity between Jupiter and the Sun is 4.10×10^{26}N

Gravitational force between Jupiter and the Sun with respect to the Law of Universal Gravitation.

From our data analysis we know that mass of the sun, m1 is 1.99×10^{30}kg, mass of Jupiter, m2 is 1898.6×10^{24}kg, mean distance between Jupiter and the sun from their centers, r is 778.4×10^{9}m and the gravitational constant, G is 6.67×10^{-11} N.m^2/kg^2 thus the gravitational force between Jupiter and the sun, F is:

By applying the law of universal gravitation,

$$\text{Gravitational Force, } F = G \frac{m1m2}{r^2}$$

$$= \frac{6.67 \times 10^{-11} \times 1.99 \times 10^{30} \times 1898.6 \times 10^{24}}{(778.4 \times 10^{9})^2}$$

$$= \frac{25200.68738 \times 10^{43}}{605906.56 \times 10^{18}} F$$

$$= 4.1591705 \times 10^{23}\text{N} \quad F =$$

$$4.16 \times 10^{23}\text{N}$$

Hence with respect to the Law of Universal gravitation the gravitational force between Jupiter and the sun is about 4.16×10^{23}N.

On the Saturn-Sun System

As a sixth planet from the sun, Saturn in its bicorporal system with the sun is acted on by external gravitational forces produced by mercury, the ether vortex in the intra-mercurial region, Venus, Earth, Mars and Jupiter. The wave principle of gravity will determine the net gravitational force acting on Saturn in its system with the sun. The law of universal gravitation will give us the gravitational force between Saturn and the sun as if the saturn – sun system is a pure bicorporal celestial system. The variation between the wave principle of gravity and the law of the universal gravitation on the saturn-sun system will verify that it is an impure   solar   bicorporal   celestial   system.

Force of Gravity between Saturn and the sun with respect to the Wave Principle of Gravity.

From our data analysis we know that mass of the sun m1 is 1.99×10^{30}kg, mass of Saturn, m2 is 568.46×10^{24}kg mean distance between Saturn and the sun from their centers, r is 1.4268×10^{12}m, Saturn's aphelion-perihelion difference, A is 158.8×10^{9}m and the mean orbital velocity of Saturn, V is 9660 m/s, therefore the force of gravity between Saturn and the sun, F is:

Under the Wave principle of gravity,

$$\text{Force Gravity, } F = n \frac{(m1 + m2)A^2V^2}{r^3}$$

$$= \frac{1(1.99\times10^{30} + 568.46 \times 10^{24})(158.8 \times 10^{9})^2(9660)^2}{(1.4268 \times 10^{12})^3}$$

$$= \frac{10^{24}(1990000 + 568.46) \times 25217.44 \times 10^{18} \times 93315600}{2.904619857 \times 10^{36}}$$

$$= \frac{4.684166972 \times 10^{60}}{2.904619857 \times 10^{36}}$$

$$F = 1.612660934 \times 10^{24}\text{N}$$

$$F = 1.61 \times 10^{24}\text{N}$$

By the wave principle of gravity the force of gravity between Saturn and the Sun is 1.61×10^{24}N.

Gravitational Force between Saturn and the sun with respect to the law of Universal gravitation.

In accordance with our data analysis we know that mass of the sun, m1 is 1.99×10^{30} kg, mass of Saturn, m2 is 568.46×10^{24} kg, mean distance between Saturn and the sun from their centers, r is 1.4268×10^{12} m and the gravitational constant, G is 6.67×10^{-11} N.m²/kg² thus the gravitational force between Saturn and the sun, F is:
By applying the Law of universal gravitation,

$$\text{Gravitational Force, } F = G \frac{m1m2}{r^2}$$

$$= \frac{6.67 \times 10^{-11} \times 1.99 \times 10^{30} \times 568.46 \times 10^{24}}{(1.4268 \times 10^{12})^2}$$

$$= \frac{7545.340118 \times 10^{43}}{2.03575824 \times 10^{24}}$$

$$F = 3.706402838 \times 10^{22} N$$

$$F = 3.71 \times 10^{22} N$$

Hence with respect to the Law of Universal gravitation the gravitational force between Saturn and the sun is about 3.71×10^{22} N

On the Uranus-Sun System

As a seventh planet from the sun, Uranus in its bicorporal system with the sun is acted on by external gravitational forces produced by mercury, the ether vortex in the intra-mercurial region, Venus, Earth, Mars, Jupiter and Saturn. The wave principle of gravity will determine the net gravitational force acting on Uranus in its system with the sun. The law of universal gravitation will give us the gravitational force between Saturn and the sun as if the Uranus – sun system is a pure bicorporal celestial system. The variation between the wave principle of gravity and the law of the universal gravitation on the Uranus-sun system will verify that it is an impure solar bicorporal celestial system.

Gravitational Force between Uranus and the Sun with respect to the Wave Principle of Gravity

From our data analysis we know that mass of the sun m1 is 1.99×10^{30}kg, mass of Uranus, m2 is 86.83×10^{24}kg mean distance between Uranus and the sun from their centers, r is 2.8710×10^{12}m, Uranus aphelion-perihelion difference, A is 271.2×10^{9}m and the mean orbital velocity of Uranus, V is 6820 m/s, therefore the force of gravity between Uranus and the sun, F is:

Under the wave principle of gravity,

$$\text{Force Gravity, } F = \frac{n\,(m1 + m2)A^2V^2}{r^3}$$

$$= \frac{1(1.99 \times 10^{30} + 86.83 \times 10^{24})(271.2 \times 10^{9})^2(6820)^2}{(2.8710 \times 10^{12})^3}$$

$$= \frac{10^{24}(1990000 + 86.83) \times 73549.44 \times 10^{18} \times 46512400}{23.66462231 \times 10^{36}}$$

$$= \frac{6.808009378 \times 10^{60}}{2.36462231 \times 10^{37}}$$

$$F = 2.876872189 \times 10^{23}\text{N}$$

$$F = 2.88 \times 10^{23}\text{N}$$

Thus by the wave principle of gravity the force of gravity between Uranus and the Sun is 2.88×10^{23}N.

Gravitational Force between Uranus and the sun with respect to the law of Universal gravitation.

In accordance with our data analysis we know that mass of the sun, m1 is 1.99×10^{30} kg, mass of Uranus, m2 is 86.83×10^{24} kg, mean distance between Uranus and the sun from their centers, r is 2.8710×10^{12} m and the gravitational constant, G is 6.67×10^{-11} N.m^2/kg^2 thus the gravitational force between Uranus and the sun, F is:

By applying the Law of universal gravitation,

$$\text{Gravitational Force, } F = G \frac{m1m2}{r^2}$$

$$= \frac{6.67 \times 10^{-11} \times 1.99 \times 10^{30} \times 86.83 \times 10^{24}}{(2.8710 \times 10^{12})^2}$$

$$= \frac{1152.520639 \times 10^{43}}{8.242641 \times 10^{24}}$$

$$F = 139.8241946 \times 10^{21} N$$

$$F = 1.40 \times 10^{21} N$$

Hence with respect to the Law of Universal gravitation the gravitational force between Uranus and the sun is about 1.40×10^{21}N.

On the Neptune-Sun System

As a seventh planet from the sun, Neptune in its bicorporal system with the sun is acted on by external gravitational forces produced by Mercury, the ether vortex in the intra-mercurial region, Venus, Earth, Mars, Jupiter, Saturn and Uranus. The wave principle of gravity will determine the net gravitational force acting on Neptune in its system with the sun. The law of universal gravitation will give us the gravitational force between Neptune and the sun as if the Uranus – sun system is a pure bicorporal celestial system. The variation between the wave principle of gravity and the law of the universal gravitation on the Neptune-sun system will verify that it is an impure solar bicorporal celestial system.

Gravitational Force between Neptune and the Sun with respect to the Wave Principle of Gravity

From our data analysis we know that mass of the sun m1 is 1.99×10^{30} kg, mass of Uranus, m2 is 102.43×10^{24} kg mean distance between Neptune and the sun from their centers, r is 4.4983×10^{12} m, Neptune aphelion-perihelion difference, A is 77.2×10^{9} m and the mean orbital velocity of Neptune, V is 5480 m/s, therefore the force of gravity between Neptune and the sun, F is:

Under the wave principle of gravity,

$$\text{Force Gravity, } F = n \frac{(m1 + m2)\, A^2 V^2}{r^3}$$

$$= \frac{1(1.99 \times 10^{30} + 102.43 \times 10^{24})(77.2 \times 10^{9})^2 (5480)^2}{(4.4983 \times 10^{12})^3}$$

$$= \frac{10^{24}(1990000 + 102.43) \times 5959.84 \times 10^{18} \times 30030400}{91.02176401 \times 10^{36}}$$

$$= \frac{35.6181327 \times 10^{58}}{9.102176401 \times 10^{37}}$$

$$F = 3.913144629 \times 10^{21} N$$

$$F = 3.91 \times 10^{21} N$$

Thus by the wave principle of gravity the force of gravity between Neptune and the Sun is 3.91×10^{21}N.

Gravitational Force between Neptune and the Sun with respect to the Law of Universal Gravitation.

In accordance with our data analysis we know that mass of the sun, m1 is 1.99×10^{30}kg, mass of Neptune, m2 is 86.83×10^{24}kg, mean distance between Neptune and the sun from their centers, r is 2.8710×10^{12}m and the gravitational constant, G is 6.67×10^{-11} N.m^2/kg^2 thus the gravitational force between Neptune and the sun, F is:

By applying the law of universal gravitation,

$$\text{Gravitational Force, } F = \frac{G \, m1m2}{r^2}$$

$$= \frac{6.67 \times 10^{-11} \times 1.99 \times 10^{30} \times 102.43 \times 10^{24}}{(4.4983 \times 10^{12})^2}$$

$$= \frac{1359.584119 \times 10^{43}}{20.23470289 \times 10^{24}}$$

$$F = 6.719071322 \times 10^{20}N$$

$$F = 6.72 \times 10^{20}N$$

Hence with respect to the Law of Universal gravitation the gravitational force between Neptune and the sun is about 1.40×10^{20}N.

Table 3.2 Net Gravitational Force in Impure Bicorporal Celestial Systems of the Sun and its Planets.

Bicorporal Celestial Systems	Mass of a Planet	Planet-Sun Mean Distance	Net Gravitational Force
	m(kg)	r(km)	F(N)
Venus-Sun	4.869×10^{24}	108.2×10^{6}	3.78×10^{24}
Earth-Sun	5.97×10^{24}	149.6×10^{6}	1.32×10^{25}
Mars-Sun	6.42×10^{23}	227.85×10^{6}	1.76×10^{26}
Jupiter-Sun	1898.6×10^{24}	778.4×10^{6}	4.10×10^{26}
Saturn-Sun	568.46×10^{24}	1426.8×10^{6}	1.61×10^{24}
Uranus-Sun	86.83×10^{24}	2871.0×10^{6}	2.88×10^{23}
Neptune-Sun	102.43×10^{24}	4498.3×10^{6}	3.91×10^{21}

In accordance with our preceding calculations in this chapter we have seen that when the wave principle of gravity and the law of universal gravitation- are applied in the solar bicorporal celestial systems, the results vary. It is thus, this variation implies· that all bicorporal celestial systems of the sun and its planets are impure bicorporal celestial systems. Note that the only significant external gravitational force in solar bicorporal celestial systems are those due to the theoretical intra-mercurial ether vortex, and other planets existing interior to the planet of a solar bicorporal celestial system on account of the centripetal acceleration factor carried by the wave principle of gravity.

In the same manner the wave principle of gravity can be used in the process of searching for exoplanets by observing the variation between it and the law of universal gravitation in systems of stars and their respective planets. The variation between the two equations of gravity will indicate the presence of an external celestial body in the celestial system under observation, since the wave principle of gravity determines the net gravitational force between two bodies whereas the law of universal gravitation determines the force of gravity between two bodies as though the system under observation is a pure bicorporal system.

Queries and Quintessences

Note that the answers to the following queries have been quintessentialized from the Cosmic Wave Theory buried herein:

Query 1: Is the empty space really empty? If the answer is yes, undefined or no, explain in what sense it is so.

Quintessence 1. No. The empty space is not realistically empty. If it would have been empty, empty in a sense that there is no presence of any medium in any form of matter at any conceivable state, there should not have been passages of electromagnetic waves such as light and elementary particles such as neutrinos from other parts of the universe to the earth. Consequently, there should be a very tenuous medium that constitutes what we call "empty space". In accordance with the Cosmic Wave Theory this medium is the ether, a universal transparent medium of extreme tenuity and exceedingly low density that interpenetrates the whole structure of the physical universe and behaves in a wave aspect as water medium.

Query 2: Are there any similarities between water medium and the ether medium?

Quintessence 2: Yes. Initially, since water and the ether are types of media, then they both have common properties such as having an ability to allow the propagation of waves, particles, forces or effects and a potential to be displaced by other forms of matter, though according to the Cosmic Wave Theory ether can only be displaced by matter in which it is already a constituent or it will simply interpenetrate the matter. And further, with respect to the Cosmic Wave Theory the ether medium behaves in a wave aspect as water medium: Just like water waves, ether waves caused by vibrations in the ether medium travel at a fixed speed to great distances, and undergo interference and diffraction processes.

Query 3: How does gravity act at a distance?

Quintessence 3: When an object is in a medium such as the ether (empty space), it continuously displaces a certain volume of ether that is equal to its volume. As a result of the action of displacing the ether, in reaction the displaced ether attempts to return to its original position with a resultant force possessed by waves of the displaced ether acting at the center of the object. This resultant force is the force of gravity of the object. It is thus, since the object (e.g a celestial body) contains ether, then the displaced ether with a force of gravity (resultant force) of the object due to the action of displacing ether tends to compress the ether medium that is already a constituent of the object, at a distance towards the center of the object.

Query 4: What is it about gravity that which makes it the weakest of the four forces of nature?

Quintessence 4: When an object is in the ether medium (empty space), it constantly displaces a certain volume of ether that is equal to its volume. As a result of the action of displacing ether, is reaction the displaced ether attempts to return to its original position with a resultant force possessed by waves of the displaced ether acting at the center of the object. This resultant force is the force of gravity of the object. It is thus since the object (e.g a celestial body) contains ether, then the displaced ether with a force of gravity (resultant force) of the object due the action of displacing the ether tends to compress the ether medium that is already a constituent of the object, at a distance toward the center of the object. On account of the ether's minute density, only the resultant force of large amounts of the ether (like that displaced by celestial bodies such as earth) is the one which tends to be effective. Consequently, in a small scale of gravity becomes a very weak force in the physical universe, and hence in general the weakest of the four forces of nature.

Query 5: What is it about the wave principle of gravity that which makes it consistent in all systems?

Quintessence 5: It is the centripetal acceleration quantity carried by the wave principle of gravity that which enables it to determine the net force of gravity between two bodies including that due to a body (or bodies), if any, that exist between the bodies of bicorporal system in question.

Query 6: In a wave perspective how does the sun influence its planets gravitationally?

Quintessence 6: The sun constantly displaces a volume of ether of about $1.4 \times 10^{27} \text{m}^3$, which is equal to its volume. As a result of displacing the ether, in reaction the displaced ether attempts to return to its original position by acting at the center of the sun with a resultant force possessed by waves of the displaced ether. This resultant force is the force of gravity of the sun. Thus, since the sun contains ether, then the displaced ether with a force of gravity of the sun (due to the action of displacing the ether) tends to compress the ether medium that is already a constituent of the sun, at a distance toward the center of the sun. it follows that the planets which are nearby or within the region covered by the ether medium displaced by the sun tend to be influenced by the force of gravity of the sun carried by waves of the ether medium displaced by it that acts toward its center. However, on account of the planets masses, they independently also have forces of gravity carried by waves of the ether medium displaced by them, which similarly act at their centers. Hence as a result of the pulling of their gravitational forces against that of the sun separately, they orbit about the sun, since their individual gravitational forces are less great than the sun.

Query 7: We know that general theory of relativity dictates that gravity is not a force but a consequence. In a wave perspective, is gravity a force, consequence or both?
Quintessence 7: In a wave perspective gravity is both a consequence and force. It is a consequence that occurs when a body in a medium constantly displaces a certain volume of the ether medium that equals its volume and a resultant force of the displaced ether medium due to the body's action of displacing it., carried by waves of the displaced ether as the displaced ether medium attempts to return to its original position by acting at the center of the body.

Query 8: We certainly know that the most indispensable part of a scientific theory is its basis. Albert Einstein's general theory of relativity basis is its famous "thin rubber sheet analogy or thought experiment". What is it about the basis of the general theory of relativity that is actually in contrary to how nature behaves?

Quintessence 8: When you imagine that space is structurally spongy or elastic and flat like a thin rubber sheet such that when a body of certain significant mass is placed on it creates a curvature of which depth is directly proportional to its mass and when another body of smaller mass is placed nearby that body consequently moves along geodesics (shortest distance between two points) in curved space time toward it which is what Albert Einstein thought of space to be like and gravity as the consequence of the curvature, then you are essentially making your imagination or perhaps I should say contemplation as if you are on earth where everything is constantly influenced by the earth's gravity and thus tends to accelerate toward its center (downwardly). In space there is realistically no up or down. Indeed no specific or conventional downward direction to which bodies should exert their weights as it held in the rubber sheet experiment. In fact, the concept of weight should completely not be incorporated in our thought experiments or analogies when we contemplate on the nature of space and gravity because bodies are weightless in space (considering the fact that weight is the amount of force of gravity due to a larger body acting on a smaller body when a smaller body is on the surface of the larger body). It is thus firstly the rubber sheet analogy or thought experiment is inaccurate and inappropriate by incorporating the weight quantity. And secondly the so called space time curvature or warp, with a depth pointing downward is not existent in nature, it is a mere conceptual error, since bodies are weightless in space. Only mass, a natural property of each body in the physical universe which is a quantity of matter in a body, is a quantity that is of a great role for bodies in space, and observing the nature of space and gravity. Most importantly, mass of each body in the physical universe acts at its center: it follows that to correct the concept of space and gravity of the general theory of relativity explicitly space is neither flat nor curved, instead it contains a tenuous, transparent and universal medium of exceedingly low density that to a great extent behaves as water medium. This medium is the ether. So when a body is placed in space displace a certain volume of the ether medium that equals its volume, as a reaction of the action of displacing the ether the displaced ether attains a resultant force that is possessed by waves of the displaced ether and acts towards the center of the body, (as the displaced ether attempts to return to its original position by acting at the center of the body). This resultant force is the force of gravity of the body since it is a force and consequence due to the presence of the body in space and its constant action of displacing the ether.

GLOSSARY

Acceleration: the rate of change of velocity with respect to time. It is a vector quantity of which SI unit is meter per second squared (m/s^2).

Andromeda: the closest galaxy of which structure is similar to that of ours. It appears in the constellation Andromeda, and is roughly 2 million light years from our Milky Way Galaxy.

Aphelion: the farthest point from the sun, on the orbit of a celestial body revolving around it.

Apogee: the farthest point from the earth, on the orbit of the natural satellite revolving around it.

Archimedes' Principle: the buoyant force on an object totally or partially submerged in a fluid is equal to the weight of the fluid displaced by it.

Atom: a tiny entity of ordinary matter made of a nucleus (composed by protons and neutrons) encompassed by electrons. It is about

10^{-10}n in diameter.

Binary Star: a member of a system of two stars revolving around their common center of mass.

Black Hole: a space-time point at which mass is so condensed, by extreme gravity, to the extent that even light cannot escape from it. This is based on the general theory of relativity.

Buoyant Force: a force exerting on an object partially or totally submerged in a fluid, due to the fluid pressure.

Centripetal Acceleration: the rate of change of uniform velocity with respect to time of a body moving in a circle and in a radial manner directed toward the center of it. It is also called radial acceleration.

Cluster: a group of galaxies consisting of a few to thousands of galaxies bound together by gravity.

Crest: the highest point of a wave alternating during the passage of a complete cycle.

Dark Matter: non- luminous matter of an unconventional form detectable by its gravitational effects on observable matter.

Density: mass per unit volume of a substance. Its SI unit is kilogram per meter cubed (kg/m^3).

Diffraction: the phenomenon in which waves bend around obstacles somewhat and pass into the region behind them.

Elementary Particle: a particle that is considered to be having a quality of not being able to be chemically decomposed into simpler particles.

Energy: the capacity to do work. It has a quality of being conserved. Its SI unit is joule (J).

Escape Velocity: the speed a body must attain so as to transcend the gravitational field of another body without a possibility of returning to it.

Ether: the universal transparent medium of exceedingly high tenuity interpenetrating all matter and behaves in a wave aspect as water medium. This is with respect to the cosmic wave theory.

Force: the push or pull exerted on a body. It's SI unit is Newton (N).

Frequency: the number of complete cycles or revolutions per second. Its SI unit is hertz (Hz).

Galaxy: a large group of stars, gas and dust bound together by gravity. It contains stars between a million and hundreds of billions.

Gamma Ray: electromagnetic waves of very high frequency and short wavelength, 10^{-12}m shorter than that of x-rays.

General Theory of Relativity: Albert Einstein's theory based on the idea that gravity is the consequence of a four dimensional space-time curvature.

Gravitational Acceleration: the acceleration of the influenced body toward the influencing body due to the influencing body's gravitational influence upon it.

Gravitational Amplitude: the greatest height of a crest or depth of a trough of the resultant wave formed due to gravitational interaction of two bodies in a medium. In a system free of external forces it is equal to the greatest displacement of an influenced body. Its SI unit is meter (m).

Gravity: the resultant force of ether medium displaced by a body, which tends to return to its original position by acting at the center of the body. This is according to the cosmic wave theory.

Heliocentric: having the center on the sun.

Impure Bicorporal Celestial System: a system of two celestial bodies in which the region between them is consisting of another celestial body (or bodies).

Index of Refraction: the ratio of the speed of light in vacuum (empty space of ether) to the speed of light in a material in question. It is denoted as n.

Influenced Body: a body (in a system) affected by another body or bodies gravitationally.

Influencing Body: a body (in a system) affecting another body or bodies gravitationally.

Intensity: the energy per unit time carried across unit area perpendicular to the direction of energy flow. It's SI unit is Watt per meter squared.

Lagrangian Point: a point in space at which a smaller body under the influence of two large ones will remain approximately at rest relative to them.

Law of Universal Gravitation: Newton's Law of Gravity based on the idea that every object in the universe attracts every other object with a force that is directly proportional to the product of their masses and inversely proportional to the square of the distance between them from their centers.

Mass: the quantity of matter in a body. Its SI unit is kilogram (kg).

Mercury's Perihelion Advance: the slight anomalous drifting of Mercury's perihelion position. It is considered to be at a rate of 43 arc sec per century. Sometimes it is called Mercury's Perihelion Precession.

Molecule: the smallest entity of a compound made by the electrical combination of two or more atoms. It exhibits the chemical properties of the compound in which it is a component.

Neutrino: a neutral elementary matter particle with exceedingly small mass released from nuclear fusion reactions and travels at a speed of light. It is only affected by weak nuclear force and gravity.

Nucleus: the core of an atom consisting of protons and neutrons in it that are held together by the strong nuclear force.

Perigee: the shortest point from the earth, on the orbit of its natural satellite revolving around it.

Principle of Superposition: in the region where wave interferences occur the resultant displacement is equal to the algebraic sum of individual displacements of the waves, whereby a crest is taken as positive and a trough negative.

Proton: a positively charged matter particle that composes the nucleus of an atom. Its mass is 1.6726×10^{-27} kg.

Pure Bicorporal Celestial System: a system of two celestial bodies in which the region between them is totally free of other celestial bodies.

Simple Harmonic Motion: motion attained by a vibrating object when the restoring force is directly proportional to its displacement.

Sinusoidal: formed in or related to sine or cosine curve whose equations in the Cartesian coordinates are of the form y=asinx and y=acosx respectively.

Special Theory of Relativity: Albert Einstein's theory based on two principles: the relativity principle which states that the laws of science are the same for all freely moving observers regardless of their speed; and the constancy of the speed of light which states that the speed of light in vacuum is constant in all inertial reference frames.

Strong Nuclear Force: one of the four fundamental forces of nature that is the strongest of all and has the shortest range. It is responsible for holding together quarks in protons and neutrons, and for holding protons and neutrons together to form nuclei of atoms.

Thermonuclear Fusion: the process in which nuclei of less heavier elements such as hydrogen combine to form heavier nuclei under extraordinarily high temperature and pressure, associated with the release of tremendously high energy. It is thought of as being the source of the sun's and other stars' energy.

Trough: the lowest point of a wave alternating during the passage of a complete cycle.

Vacuum: the empty space formed after the removal of air medium. In the cosmic wave theory it is considered as the ether.

Velocity: the rate of change of distance with respect to time in a specified direction. Its SI unit is meter per second (m/s).

Vulcan: a hypothetical planet predicted to exist in the region interior to mercury. According to the cosmic wave theory the theoretical ether vortex in the intra-mercurial area takes the place of Vulcan.

Vulcanoids: hypothetical planetoids predicted to exist in the intra-mercurial region.

Wavelength: the distance between two successive crests or two successive troughs of a continuous sinusoidal wave.

Wave Principle of Gravity: a scientific principle based on the idea that the force of gravity between two bodies in any medium is directly proportional to the product of the sum of their masses, the square of the gravitational amplitude of their resultant wave, and the acceleration of the influenced body but inversely proportional to the square of the mean distance between them from their centers.

Weak Nuclear Force: one of the four fundamental forces of nature that is the second weakest of all and has a very short range. It affects matter particles and not force carrying particles, and is responsible in radioactive decay.

AUTHOR'S NOTE

<u>ON THE APPLICATION OF THE COSMIC WAVE THEORY</u>

Dear readers as we have seen, this first edition of the cosmic wave theory advances our understanding on the structural nature and behaviors of the ether or space or vacuum and the force of gravity. In the **second edition of the cosmic wave theory** and in the new branch of physics called **INFINITUM PHYSICS** given birth to by this cosmic wave theory, we are going to advance our understanding on the structural nature and behaviors of black holes and dark matter in relation to the formation of ether vortices and fusion energy. The following are practical applications of the cosmic wave theory which are exponentially ex post facto, just to mention a few which will be fully explained in **INFINITUM PHYSICS** as well.

<u>In Astronomy</u>

As I have said before, the wave principle of gravity buried herein can be used in the process of searching for exoplanets by observing the difference between it and the law of universal gravitation in systems of stars and their respective planets. The significant variation between the two equations of gravity will indicate the presence of an external celestial body in the system under observation, since the wave principle of gravity determines the net force of gravity between two bodies whereas the law of universal gravitation determines the force of gravity between two bodies as though the system is a pure bicorporal celestial system.

In Fusion Energy Technology

There are sine qua non derivations and explanations which will be laid in **INFINITUM PHYSICS** that are integral keys in the process of practicalizing thermonuclear fusion energy, in other words simulating a small scale sun –a power source that has capacity of providing an inexhaustible supply of energy without global warming or pollution.

PHILOSOPHICAL CONVICTIONS ON NATURE, SCIENCE and TECHNOLOGY

Nature is like

a balanced,

Infinitely complex

matrix.

To be in tune with the hidden behaviors of nature, one

must philosophically

Or mathematically penetrate

Through the subtle fringes of the ether's texture.

It is in the infinite ether's well,

Whereby magi, bards, vates and philosophers dwell.

Thereby they become aware that they are a mere part of nature.

And thus nature is conscious of her perpetual existence

through their minds.

Indeed, through the ultra ether in our brains

We are aware of our awareness, and aware of the fact
that

Of nature we are a mere part.

Science is an exact

And consistent set of knowledge That

enables humankind to define nature.

A homogeneous blend

Of mathematics and philosophy

Is a sine qua non key to discovering behaviors of nature

unknown to man and hence advance science.

What makes a scientific theory eternal

Is its equation's exactness,

And the transcendental importance
Of its philosophy and equation.

A civilization's scientific advancement

Can purely be determined by measuring its

development in mathematics, philosophy and

science;

Whereas technology is merely the trio's consequence.

A scientific advancement of a civilization

Ought to lead to the increase in its spiritual aperture,

On account of a new realization and awakening

That comes with discovering unknown behaviors of nature.

Technology is the implementation and manipulation

of science and mathematics

In creating techniques

For the benefit of humanity.

To prevent our civilization's sole seizure,

We ought to make sure,

Each new technology we introduce

Is congenial with nature.

----END---

Stay in tune for the **second edition**

of the COSMIC WAVE THEORY

and the new branch of physics

delivered by the cosmic wave

theory: **INFINITUM PHYSICS;**

the physics of things that move

faster than light and their

mediums.

Says

Francis Mlaga.

Author, Scientist, Physicist, Mathematician and Engineer

www.ingramcontent.com/pod-product-compliance
Lightning Source LLC
Chambersburg PA
CBHW020921180526
45163CB00007B/2835